I0750912

REPORT OF A COMMITTEE

OF THE

BOSTON BOARD OF TRADE

UPON THE

COTTON TAX.

BOSTON:
PRINTED BY PRENTISS & DELAND,
40 CONGRESS STREET.
1867.

THE

PROPOSED REPEAL OF THE COTTON TAX.

BOSTON, Dec. 26, 1867.

HON. HENRY WILSON AND HON. CHARLES SUMNER:

Gentlemen, — The undersigned, a committee appointed by the Boston Board of Trade to consider the question of the tax upon cotton, have the honor to present to your notice the accompanying report, made by them to the board; which report was accepted, and the vote therein recommended was adopted: —

REPORT.

The Committee of the Boston Board of Trade, to whom has been referred the question of what action the Board should take in reference to the repeal of the tax now imposed upon American cotton, report the following vote, and recommend its adoption: —

Voted, That our Senators be requested to vote for the bill which has passed the House of Representatives and is now before the Senate, by which the tax now imposed on American cotton will be removed.

They recommend the adoption of this vote for the following reasons: —

1st, Because the following statements of the consumption of cotton in Europe, and estimates of the supply for one year from Oct. 1, 1867, are fully credited in Europe; and, until the estimates of supply are discredited by the facts, the price of cotton must rule very low as compared with the price prevailing when the tax upon cotton was imposed.

Your Committee are well aware of the danger of basing the probable course of prices upon statistics; but even those who most distrust statistics will admit, that, whatever the facts may prove, what people believe to be facts will affect their action, and consequently affect prices.

CONSUMPTION OF COTTON.

OCT. 1, 1866, TO OCT. 1, 1867.

	AMERICA.	INDIA.	BRAZIL.	EGYPT.	SUNDRIES.	TOTAL.
ENGLAND.	Bales.	Bales.	Bales.	Bales.	Bales.	Bales.
Stock in ports, Oct. 1, 1866	268,000	515,000	96,000	24,000	42,000	945,000
Imported Oct. 1, 1866, to Oct. 1, 1867	1,222,000	1,444,000	417,000	181,000	153,000	3,417,000
	1,490,000	1,959,000	513,000	205,000	195,000	4,362,000
Exported	230,000	678,000	88,000	10,000	31,000	1,037,000
	1,260,000	1,281,000	425,000	195,000	164,000	3,325,000
Stock, Oct. 1, 1867, according to brokers	244,000	466,000	127,000	35,000	39,000	911,000
Consumption	1,016,000	815,000	298,000	160,000	125,000	2,414,000
CONTINENT.						
Stock, Oct. 1, 1866	82,000	71,000	21,000	2,000	22,000	198,000
Direct Imports	276,000	80,000	69,000	47,000	225,000	697,000
Exports from England less 18,000 sent *to* England	227,000	678,000	83,000	10,000	21,000	1,019,000
	585,000	829,000	173,000	59,000	268,000	1,914,000
Stock, *Sept.* 30, 1867. — Havre, 95,000 bales; Hamburg, 25,000 do.; Bremen, 21,000 do.; Sundries, 40,000 do.	53,000	52,000	21,000	4,000	51,000	181,000
Consumption	532,000	777,000	152,000	55,000	217,000	1,733,000
Total	1,548,000	1,592,000	450,000	215,000	342,000	4,147,000

Within about 10 per cent in weight of the consumption in 1860.

ESTIMATE OF SUPPLY OF COTTON TO EUROPE.

Oct. 1, 1867, to Oct. 1, 1868.

	Bales.
From America	1,750,000
From Egypt	250,000
From Brazil and other countries (supply of 1865–6, 915,000 bales; 1866–7, 849,000 bales), say average 1867–8	900,000
From India (supply of 1865–6, 1,992,000 bales; 1866–7, 1,524,000 bales), say average 1867–8	1,750,000
	4,650,000
Add stock in Europe, Oct. 1, 1867	1,092,000
	5,742,000
Consumption in 1867–8 (50,000 bales in England, 33,000 bales on Continent), per week 83,000	4,342,000
Stock, Oct. 1, 1868	1,400,000

Or 300,000 bales more than Oct. 1, 1867.

Average Weight. — American, 450 lbs.; Brazilian, 175 lbs.; Egyptian, 490 lbs.; East Indian (Surat or Bombay, 383 lbs.; Madras or Bengal, 300), 340 lbs.; Sundries, 300 lbs.

Such a supply may not come forward in consequence of low prices, but there is sufficient evidence that crops which would warrant such a supply have been made; and it is only a question of price whether it comes forward or not. There can be but little doubt that a supply even much less would be ample for the probable demand, as the power of Europe to consume cotton goods during the ensuing year has been very much curtailed by the disturbed condition of political affairs, but far more by bad harvests and the high prices of food. The great consumers of cotton goods are the mass of the people who live upon the earnings of each year and have little or no surplus; and the more they have to spend for food the less they have for clothing. To a certain extent, cotton goods have also been superseded by other fabrics during the period of high prices.

2d, Your Committee urge the repeal of the tax upon cotton, because it indirectly gives too great encouragement to the growth of cotton in other countries. How great this encouragement is may be realized from a statement of the cost of raising cotton in India, our principal competitor.

Upon this point your Committee are convinced, from the evidence that has been placed before them, that the cost of raising cotton in India, allowing only forty pounds as the product of clean cotton per acre, does not exceed three pence sterling, — say six cents per pound, — and that such cotton can now be laid down at 4¼ pence — say 8½ cents per pound — in Liverpool.

The quality of India cotton has been much improved, partly from better care in the selection of native seed and the use of exotic seed, but mainly from better processes in ginning and packing, — especially in ginning; in which department new and improved gins have been introduced to a very great extent, and of a model which we may find it expedient to adopt in this country.

It may be said that India cotton is now worth on the average seventy per cent the price of middling uplands; while a very considerable portion of the crop is worth eighty to one hundred per cent for the manufacture of coarse goods. It will be observed that the consumption of India cotton upon the Continent of Europe was last year much greater than that of American; while in England it was less, — the reason being that the goods made upon the Continent are coarser, and labor is much cheaper.

India cotton did not cease to be grown and shipped to a certain extent when the average price in Liverpool ranged from three to four pence per pound, and the cost of transportation was far more than it now is. If it be admitted that it can now be laid down in England at 4¼ pence, or 8½ cents, and is worth on the average only three-fourths the price of American, then, in order to regain the control of the cotton market of the world, we must be able to lay down middling uplands in Liverpool at 5¾ pence, or 11½ cents, all taxes and charges paid.

It is true that our tax of 2½ cents is collected in currency; but, with the cost of collection, the interest on the additional capital required to move the cotton, and other incidental expenses, it is equal to 2 cents in gold, or 1 penny. Deduct this from 5¾ and we have 4¾ pence, or 9½ cents, left to cover the cost of raising our cotton, and sending it to market. It may well be asked whether we can at present do that; and this leads us to the third point upon which your Committee base their opposition to the tax upon cotton; viz., that the agricultural system of the South is undergoing an entire revolution.

Cotton did not cease to be grown at the South, nor did the crop cease to increase, even when middling cotton fell to 4 cents per pound in Southern markets; nor did the average range of 9¼ cents in Liverpool for the years 1843, '4, and '5 prevent a very rapid increase in our crop; thereby proving that it was then a profitable crop, even under the wasteful and costly system of slave labor.

Your Committee have entire confidence in the ultimate cheapness and economy of free labor as compared with slave labor, and they believe that this country, with its superior climate and soil, will presently supply as large a proportion of the demand for cotton, at as low prices, and with more profit than formerly, unless, as education and skill increase, more valuable commodities shall drive out cotton, in which case our loss would be our gain; but, during the change from the plantation to the farm system, the cost of cultivation must be high, and it is during this period that every impediment should be removed.

When slavery was abolished, the plantation system was doomed. The improved lands of the South have heretofore been held in large parcels, a small portion only under cultivation, upon an exhaustive plan, without any rotation of crops, — the remainder held for the purpose of keeping a free population at a distance, and to supply new fields as the old ones became exhausted. The change in the social order must involve an entire change in the holding of land: it will no longer be for the interest of the owner to repel immigration, but to invite it: his fictitious capital in slaves having been destroyed, he must now find real capital or value in his land; he must sell a portion in order to make the rest more valuable; he must invite the farmer to come to his aid, — the man who will get large crops from a small number of acres; he must use good tools and implements, and endeavor to introduce agricultural machinery, as in the West; and, to accomplish this, he must educate the laborer. The slave could only be trusted with rude and heavy tools; and, however intelligent the planter might have been, — and we do not deny that very many of them were very intelligent, — yet they were crippled by the very ignorance which the necessity of their system obliged them to enforce by law.

But the change is in process; and, without confiscation, and without incurring the great risk of ruining the negro population by bestowing land upon them before they have, by earning it, educated themselves to its proper use, the natural and beneficent law of freedom is working out its logical result; and we may expect, before many years, to see such changes in the social order and land tenure of the South as will bring that section to harmony and real union with the rest of the country.

But to such small farmers, — to the men who by their own labor cultivate their little patch of cotton, and, with their wives and children, pick it, — the imposition of a tax of $10 to $12 per bale, or 20 per cent upon its present value, is a most onerous burden. The laboring man who has five or ten bales of cotton from his little patch cannot be expected to have $60 to $100 on hand in money before he has sold it, and he must either borrow money at high rates or sell his cotton at a disadvantage.

Your Committee believe that the South must continue to raise cotton, as its saleable product, to a very large extent. Until it secures a more dense and a better-educated population, its crops must be such as can be raised by what may be called the ruder or simpler methods of cultivation, and such crops are grain and cotton. And, while the Southern States will doubtless raise more grain for their own consumption than they have before, it is not believed that they can, to any very great extent, produce grain to sell in competition with the better lands and far better methods of culture of the Western States: hence their only alternative for a saleable commodity must be cotton, for many years to come.

Your Committee regret that the tax upon cotton had not been removed by the passage of an act in February last, to take effect Sept. 1, 1867. At that time two members of your Committee were in Washington, and earnestly seconded the effort of Hon. James G. Blaine, of Maine, to secure the passage of a bill which he then introduced, which was adopted in committee of the whole, but thrown out by a close vote in the House. The relief is imperatively demanded at the present time; and your Committee would have advocated the passage of a law to take effect at once, had it not appeared to be impracticable to take such action in the House of Representatives.

As long since as April, 1866, one of your Committee (while

advocating a tax not exceeding 2 cents, as a temporary measure, and opposing the attempt then being made to fix the rate at 5 to 10 cents per pound) used the following language: —

"Given a supply of 5,000,000 bales, *or barely enough*, is there not danger that the average price, free from taxes, would not exceed 20 cents; say for Surats or short staples, 17½; American, 22? These being the average values, in gold, without regard to taxes, and the supply being barely sufficient, the attempt to add the tax of 5 cents would be partially successful; and it would probably result in enhancing the price, say to 19 or 20 for Surats, and 24 to 25 for American.

"But let us look forward a single year beyond. Let it be admitted that the supply delivered from Nov. 1, 1866, to Nov. 1, 1867, shall be less than 5,000,000 bales, less than *enough*, and the price consequently so high as to enable this country to add the five cents' tax, can any one doubt that such price would still further stimulate production; carry that of other countries nearly or quite to 3,000,000 bales, and our own crop to 3,000,000 bales or over? Then the price must fall to a low point, and the only tax which could possibly be borne by American cotton would be one which should represent less than the difference in value between American and Surat cotton, estimating such values at old-fashioned, or what may be called normal prices, say eight to nine cents for Surats, and ten to thirteen cents for American. In such event, two cents would be the highest point, and even that would have to be temporarily removed.

"It may seem absurd to intimate even the possibility of such prices so soon as the year 1868; but the whole question turns on the aggregate crop of the world being a *little more than enough*, and, if 6,000,000 bales be a little more than enough, such prices are possible.

"It may not be denied that at such reduction in price the cultivation of cotton would cease in many parts of the world; but India and Egypt would be slow to give up the struggle, and, in order to regain the monopoly or absolute control of the markets of the world, — which the writer fully believes our great superiority, both in point of soil and climate, entitle us to, — our cotton must be absolutely free from tax during the period of low prices which must inevitably follow the excessive prices which have prevailed."

We have given upon a previous page the estimated supply and stock of Europe for one year, from Oct. 1, 1867, to Oct. 1, 1868 . . . . 5,742,000 bales.

Let us add, for our own stock and consumption, the very moderate quantity of . . . . 800,000 bales.

And we have . . . . 6,542,000 bales.
or a possibility of a good deal more than enough. The result has followed: the last quotation in Galveston for low middling Texas cotton, a quality equal to middling uplands, was ten cents per pound in gold.

Respectfully submitted,

E. R. MUDGE,
GEO. L. WARD,
EDWARD ATKINSON,
C. W. FREELAND,
C. O. WHITMORE,
} *Committee.*

DECEMBER, 1867.

We observe by the reports of the debate in the Senate, that our allegation of the comparative value of India and American cotton has been questioned; but the doubt seems to have been induced upon the statement of some one who must be practically ignorant upon the subject, as it is based upon a comparison of "middling Surat" with "middling uplands." The simple answer to this statement is, that Surat, or India cotton is classed or graded upon an entirely different system from our own. A low grade of India, corresponding to our "ordinary," is called "middling," while the grade corresponding to our "middling" is called "fair to good."

On the 5th December, the prices were quoted as follows, in Liverpool: —

"Middling" upland . . . . . . $7\frac{1}{4}$ to $7\frac{1}{2}$ pence.
"Fair to good" Broach, Dhollerah, Oomrawuttee, &c., being India cottons . . . . 6 to $6\frac{1}{2}$ pence.

But, to set this question at rest, we now present the following letters from two of our best practical manufacturers, who have passed several months, during the past year, in Europe: —

29 Kilby St., Boston, Dec. 20, 1867.

Dear Sirs, — Your favor of the 18th inst., relative to India cotton, is received; and in reply I have to say, that, having spent the early part of 1864 in the manufacturing districts of England, and having just returned from a ten months' tour in England and on the Continent, I will give you the result of my observations and investigations relative to the matter in question.

While in England, in 1864, I examined in detail the machinery which had then been altered and adapted to the manufacture of Surat cotton. This was then giving manufacturers a good deal of trouble, even for coarse yarns, on account of its being improperly baled, and the staple being badly cut in ginning; and the work ran so badly that the operatives were only induced to remain in the mills, where Surat cotton had been substituted for American, by the stern necessity of obtaining a livelihood. But this year, I found, to my surprise, but little objection in Europe to Surat cottons, for goods of coarse and medium numbers. Upon careful investigation, I found that this was owing somewhat to further alterations and adaptations of machinery, but more especially to the improved quality of the cotton. This improvement was due mainly to the introduction into India of the Macarthy gin, made by Messrs. Platt Brothers & Co., of Oldham, England; which has superseded nearly all others there.

In Hindostan, from time immemorial, a rude handmill, called a "churka," has been used for this purpose. It consists of a rude framework, bearing two rollers of teak wood, fluted lengthways by five or six grooves, and revolving nearly in contact. The cotton, as it is drawn between the rollers, is freed from the seeds, they being too large to pass through.

Prior to our war, much of the cotton in that country was ginned in this primitive machine; which, I suppose, has remained for more than two thousand years without alteration or improvement. With this machine, twenty persons could produce only one hundred pounds of clean cotton in one day; while with the Macarthy hand gin, twelve inches wide, one man will produce eight pounds per hour. The same machine, forty inches wide, driven by water or steam power, will produce thirty pounds per hour, or as much in ten hours as sixty persons produced on the "churka" in the same time. The Macarthy gin takes the cotton from the seed as clean, and with as little injury to the staple, as can be done by human fingers; thus saving all the staple, and leaving the seed in much better condition for planting than from the saw gin. Doubtless this gin is destined to take the place of all others in our own country, both on account of its superior qualities, and because the hand gin is especially suited to those planters who raise but few bales, and cannot afford the machinery and power required for the gin in present use. It is adapted to all staples of cotton; and one of them can be seen in use, by calling upon D. Keith, Esq., of Columbus, Georgia. For the past four years, Messrs. Platt Brothers & Co. have made and sent to India at the rate of nearly one thousand gins per month; and still the demand continues. I was assured by many manufacturers, that it is a fact that cotton put through the Macarthy gin brings 1½ pence per pound more than the same cotton put through the American saw gin.

I was everywhere met by manufacturers with this question, "How cheap are you going to produce cotton in America? For, unless you furnish it at nearly as low rates as the Surat cotton, we shall buy only your better grades for fine, and your Sea Island for very fine, numbers." It is my opinion, that, in order to control the market of Europe, the greater part of our cotton must be laid down in Liverpool at from 5 to 5½ pence, or 10 to 11 cents per pound; as the general expression and sentiment there is to that effect.

I have thus briefly given you my ideas upon this subject, and remain,

Your obedient servant,

WILLIAM H. THOMPSON.

LEWISTON, ME., Dec. 23, 1867.

DEAR SIRS, — When in England, in 1866, I devoted much time to the investigation of the question of working of Surat and other foreign cottons, keeping in view not only the *practicability* of *working* Surats into what we call fine numbers, — say from thirty to sixty, — but also the *comparative value* of the various foreign and home productions. I have worked a good deal of India cotton into coarse fabrics before going to England, and have succeeded quite well in all numbers up to twenty-five; but had never attempted any thing finer, except for filling, and by mixing with American. In England, I found them spinning all numbers from sixteen to sixty from clear Surat, and producing entirely satisfactory quality of work; and I was surprised at the degree of perfection to which they had arrived in spinning the finer as well as coarser numbers. The explanation is this: their machinery was adapted to work short-staple and dirty cotton for coarse and medium numbers; and the finer numbers were made from a quality entirely different, and superior to any India cotton I had ever seen or conceived of,—the staple being fine and sound and nearly as long as American, good color, quite as clean as our grade of middling, and possessing as good spinning qualities as our uplands: and this was not of rare occurrence, but exceedingly common.

I then understood how the English had so much advantage over the Americans. They not only had about twenty-five per cent cheaper *labor* and *supplies* (on a gold basis), but at least twenty-five per cent cheaper *cotton* also, for all numbers which come in competition with the products of our mills. Indeed, for numbers ranging from thirty down, the difference was even *greater* than that. And this was not so much the result of superior skill, and the adaptation of their machinery to cheap cotton, as the fact that they could buy India cotton, which would answer their purpose, so much cheaper than they could American. And, if we could procure India cotton as much cheaper than we can American as the English do, there is no doubt it would be very extensively used here in the manufacture of print cloths and sheeting and drills,—from No. 12 to No. 25,—coarse yarns, bags, &c. Indeed, I should not hesitate to work the higher grades of Surats for filling into such goods as the Androscoggin L's,* Bates XX, and Hill Simple Idems.

It is a great mistake to suppose that the world is dependent upon the United States for cotton, even for fine numbers; for, notwithstanding there is a good deal of poor cotton grown in India and other countries, there is also a large amount of better

* The goods named are such as are used by the best shirt-makers. No. 25 is the average number of the manufacture of the United States. The number means so many skeins, or hanks, of 840 yards each, to one pound.

grades produced. Egyptian and South-American cottons are of very superior quality,—fully equal, if not superior, to the best grown in the United States, except Sea Island; and I found, in spinning 80's to 120's, the Egyptian was preferred to almost any other.

Yours very truly,

A. D. LOCKWOOD.

In this connection, the following estimate of raising cotton in Egypt will be interesting. It is from the report of Wm. S. Thayer, late United-States Consul-general at Alexandria, dated March 5, 1863:—

Tax paid to Government,	110	piastres
Ploughing,	50	"
Irrigation,	60	"
Seed,	20	"
Hoeing,	100	"
Picking,	100	"
Ginning,	40	"
	480	"=$24.00

"The above statement was furnished me by a successful planter at Mansaneh, in Lower Egypt; but the items are upon a scale of expense considerably larger than is necessary in some of the other districts."

"As an acre in Mansaneh yields an average of four cuntars (94 lbs. net each), the expense of raising one cuntar will amount, according to the foregoing statement, to six dollars."

Less than 6½ cts. per pound for a variety of cotton which is superior to any of our cotton, except our Sea Island, for which it is largely used as a substitute. If Egypt were under a civilized government, we should find her our most dangerous competitor.

Land of the best quality, sufficient to produce 2,000,000 to 3,000,000 bales of cotton, could be brought into cultivation in Egypt, if the requisite labor could be obtained.

We also take leave to present the following letter from one of our principal cotton States, received after our report had been presented, and which fully confirms one of the positions therein taken:—

"It is evident that the present system of raising cotton in the Southern States will have to be materially changed before large crops can again be looked for. The introduction of better farming implements and labor-saving machines may do much, but it will take time to perfect them; and it has already been demonstrated during the past two seasons that planting cotton on large plantations, with hired labor, on the old plan, will not pay; while the small farmer, who hires only one or two hands, and gives his crops the benefit of his own labor and industry, can do very well. The large land-owners are anxious to sell the greater portion of their lands, and offer to divide them into small farms at less price than the public lands are now being sold; but the great depression in all departments of industry, and feeling of uncertainty regarding the future political status of the cotton States, makes it very difficult to find purchasers, even when long credit is tendered.

"It will take time also to eradicate old-fogy notions and prejudices; but we note with pleasure that this is gradually being accomplished. It has been a source of both pride and pleasure to the writer to see hundreds of young white men, who before the war would have scorned to hold the plow-handles, now at work in the fields, endeavoring to resuscitate their fallen fortunes by their own industry and honest labor. Such cannot fail to succeed."

We send with this, for the examination of yourselves and your associates, two cases, containing the collection of samples of all known varieties of cotton referred to in the Report of the Commissioners to the Paris Exhibition. This collection was made for Mr. B. F. Nourse, of Boston; and is probably the most complete one that has ever been made.

It may here be proper for us to advert to a question of great interest to us as manufacturers and merchants; viz., the abatement of the *duty* of 3 cts. per pound in gold, now levied upon raw cotton imported.

Two of the members of this Committee have purchased, and caused to be manufactured, from 1860 to 1864, over 25,000 bales of India cotton. They therefore speak of its value from their own knowledge.

The duty is now prohibitory; and, while it is now of perhaps little importance, it may soon become imperatively necessary to us to be able to obtain foreign cotton on even terms with English manufacturers, if we expect to compete with them in other markets, in the export of cotton fabrics. Our export of heavy goods has been resumed to a very considerable extent; and many spindles are now operating upon heavy goods which would be stopped, except for the demand for China and South America.

But, with the repeal of the tax upon cotton, the draw-back now allowed upon cotton goods will, of course, cease; and our export trade may be impeded unless we can secure India cotton. To cause our exports of heavy fabrics to cease would not only be a misfortune to us but to the South and West, as we have now more coarse spindles than will supply the home demand; and the extension of cotton spinning, which we heartily desire to see, may be retarded by the competition of the mills now existing in New England, which stand at a low cost and are in skilful hands.

We may also add that the free importation of India cotton will enable us to make the seamless grain bags, now so universally used, at a much lower cost than they can otherwise be made.

Very respectfully,

GEORGE L. WARD,
EDWARD ATKINSON,
C. W. FREELAND,
E. R. MUDGE,
CHARLES O. WHITMORE, } *Committee.*

www.ingramcontent.com/pod-product-compliance
Lightning Source LLC
LaVergne TN
LVHW020637110826
845149LV00004B/1251
9781418191061